First Steps In Reading

Phyllis Bertin
Eileen Perlman

School Specialty, Inc.
Cambridge and Toronto

Table of Contents

School Specialty, Inc.

Printed in Benton Harbor, MI, in October 2018
ISBN 978-0-8388-5137-1

14 15 16 17 18 PPG 22 21 20 19 18

First Steps In Reading

First Steps In Reading is an instructional tool that bridges the gap between phonemic awareness (an understanding that words are made up of sounds) and beginning reading. It contains activities for teaching sound/symbol association, blending, word recognition, reading, spelling, and handwriting, including reading individual words, phrases, and sentences. These activities are based on validated research and reflect the best current educational practices.

First Steps In Reading was written to coordinate with *PAF (Preventing Academic Failure)*, a multisensory program for teaching reading, spelling, and handwriting in grades K–3. It can be used, however, with any beginning reading program.

A multisensory program is a way of teaching reading that integrates reading, spelling, and handwriting into unified lessons. Children learn to see the letters (visual input), say the letter sounds (auditory input), and write the letters (kinesthetic input) simultaneously. *First Steps In Reading* provides practice in these multisensory skills.

First Steps In Reading reinforces the first lessons taught in the PAF program:

- The ability to isolate initial sounds and blend sounds into words
- The sound/symbol associations for the consonants *c, t, d, g, s, f, m, l, h, p,* and *n*, and the short vowel *a*
- The ability to read and spell all consonant-vowel-consonant words composed of these letters
- The lowercase motor patterns for writing these letters

PAF (Preventing Academic Failure): A Multisensory Curriculum for Teaching Reading, Spelling, and Handwriting, by Phyllis Bertin and Eileen Perlman, Educators Publishing Service, Cambridge, MA

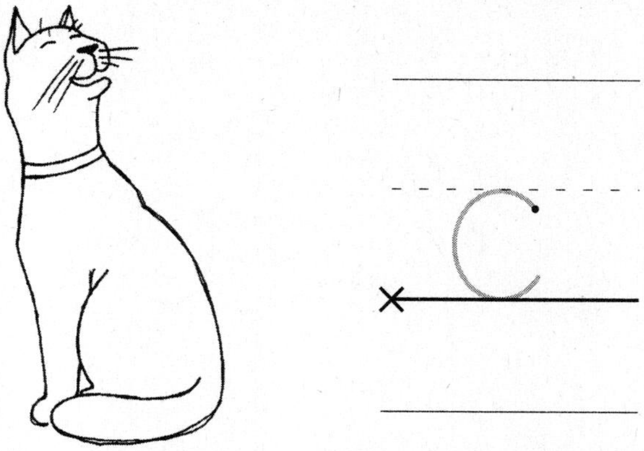

Color the pictures that begin with the sound /*k*/.

c	c
c	c

Draw pictures that begin with the sound /*k*/.

c

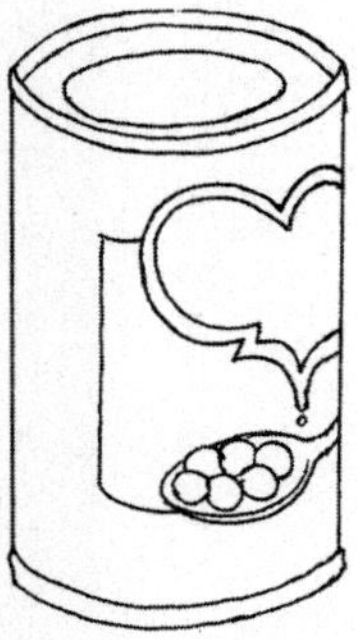

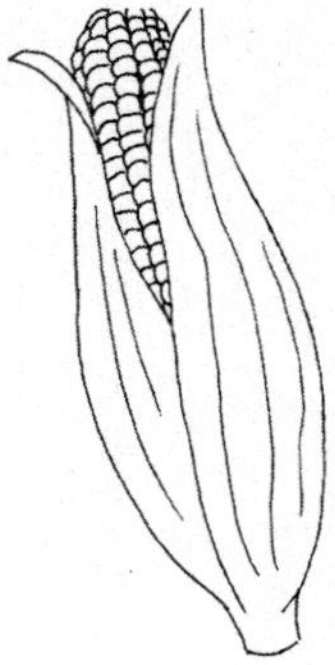

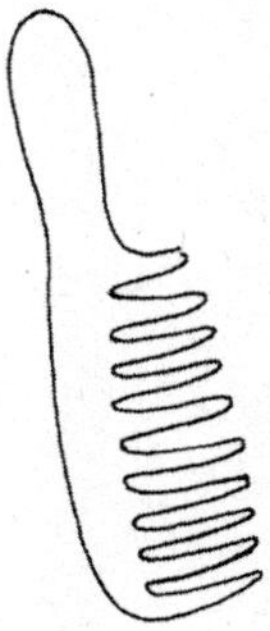

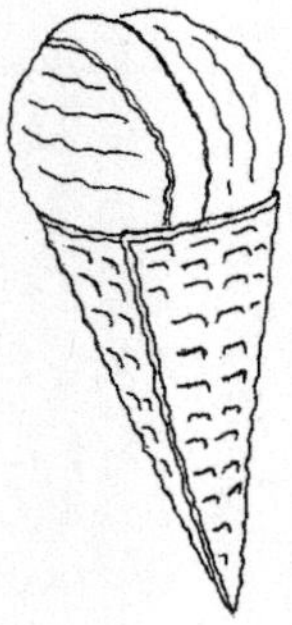

Write the letter *c* next to each picture that begins with the sound /*k*/.

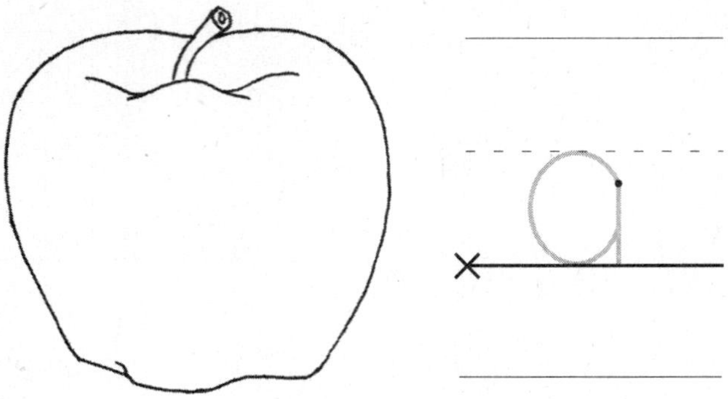

Color the pictures that begin with the sound /ă/.

c	a
a	c

Draw pictures that begin with each sound.

a

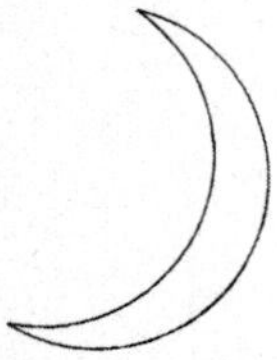

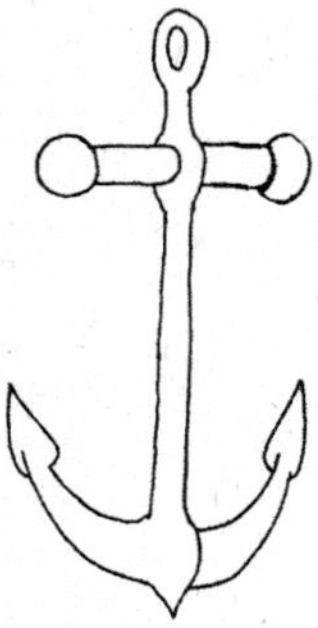

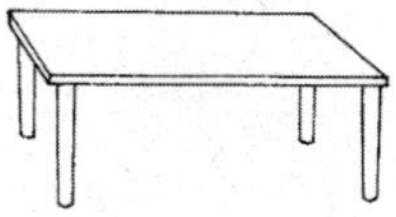

Write the letter *a* next to each picture that begins with the sound /ă/.

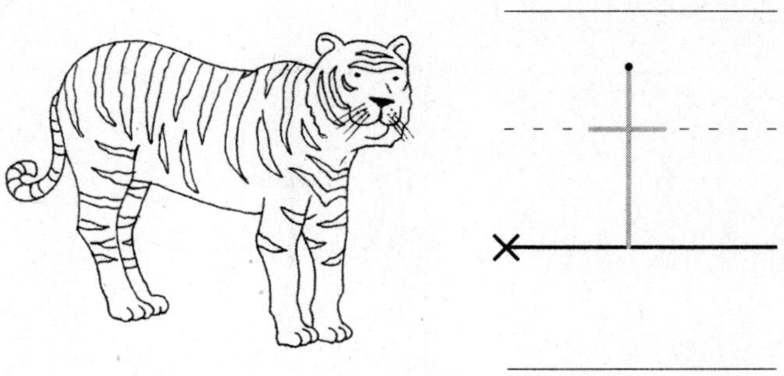

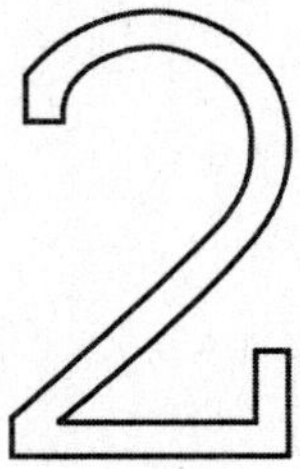

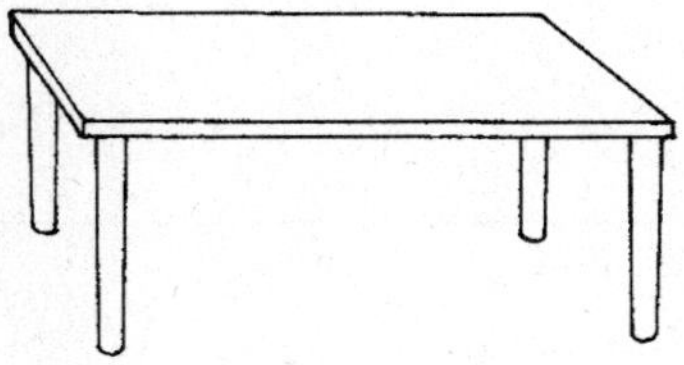

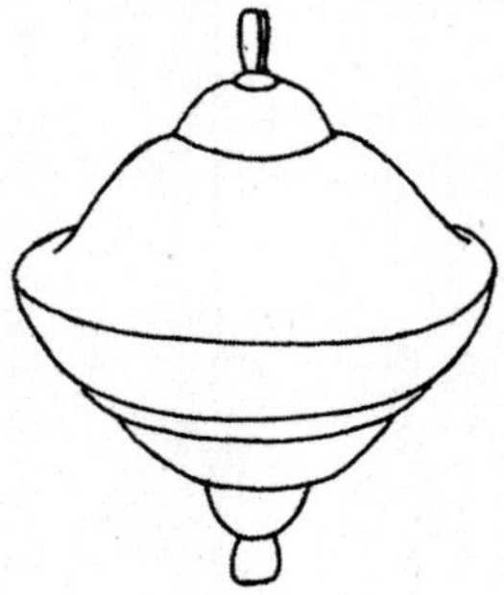

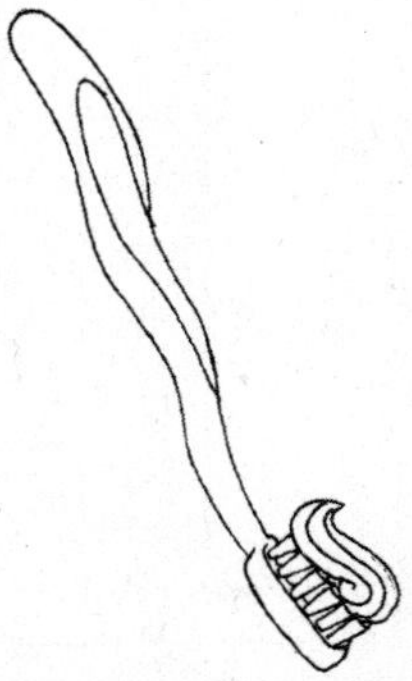

Color the pictures that begin with the sound /*t*/.

c	a
a	t

Draw pictures that begin with each sound.

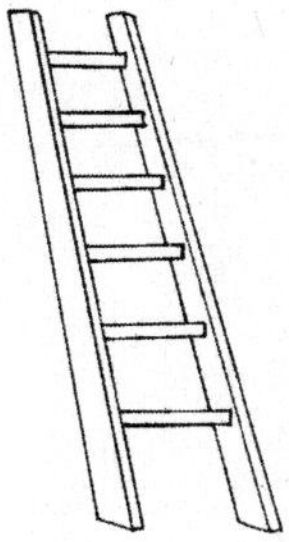

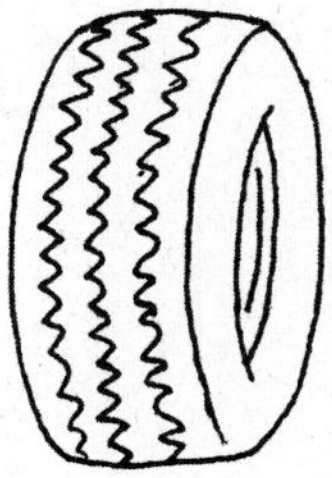

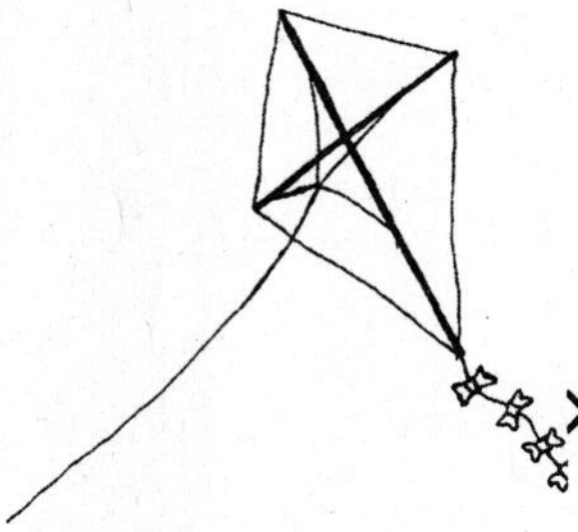

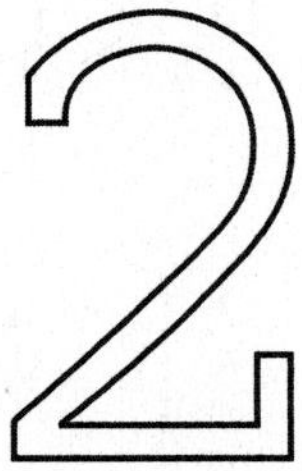

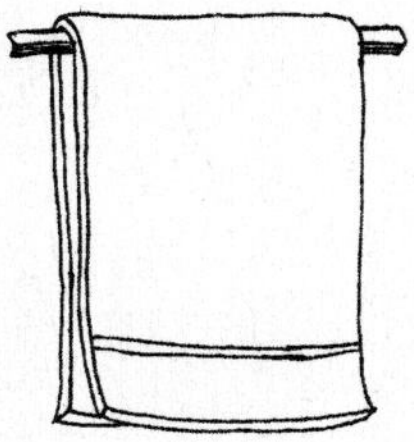

Write the letter *t* next to each picture that begins with the sound /*t*/.

cat

Read the word *"cat."* Trace and copy the word.

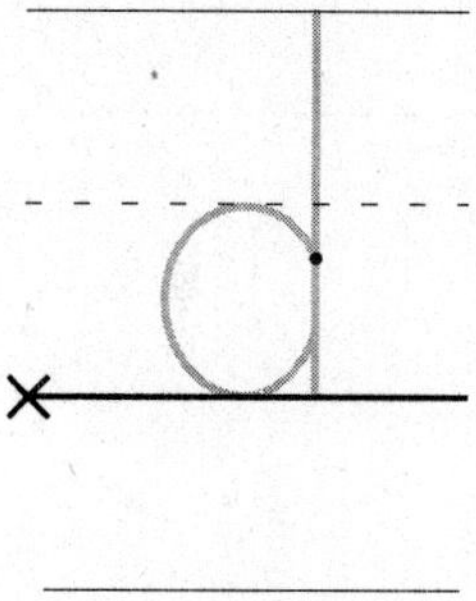

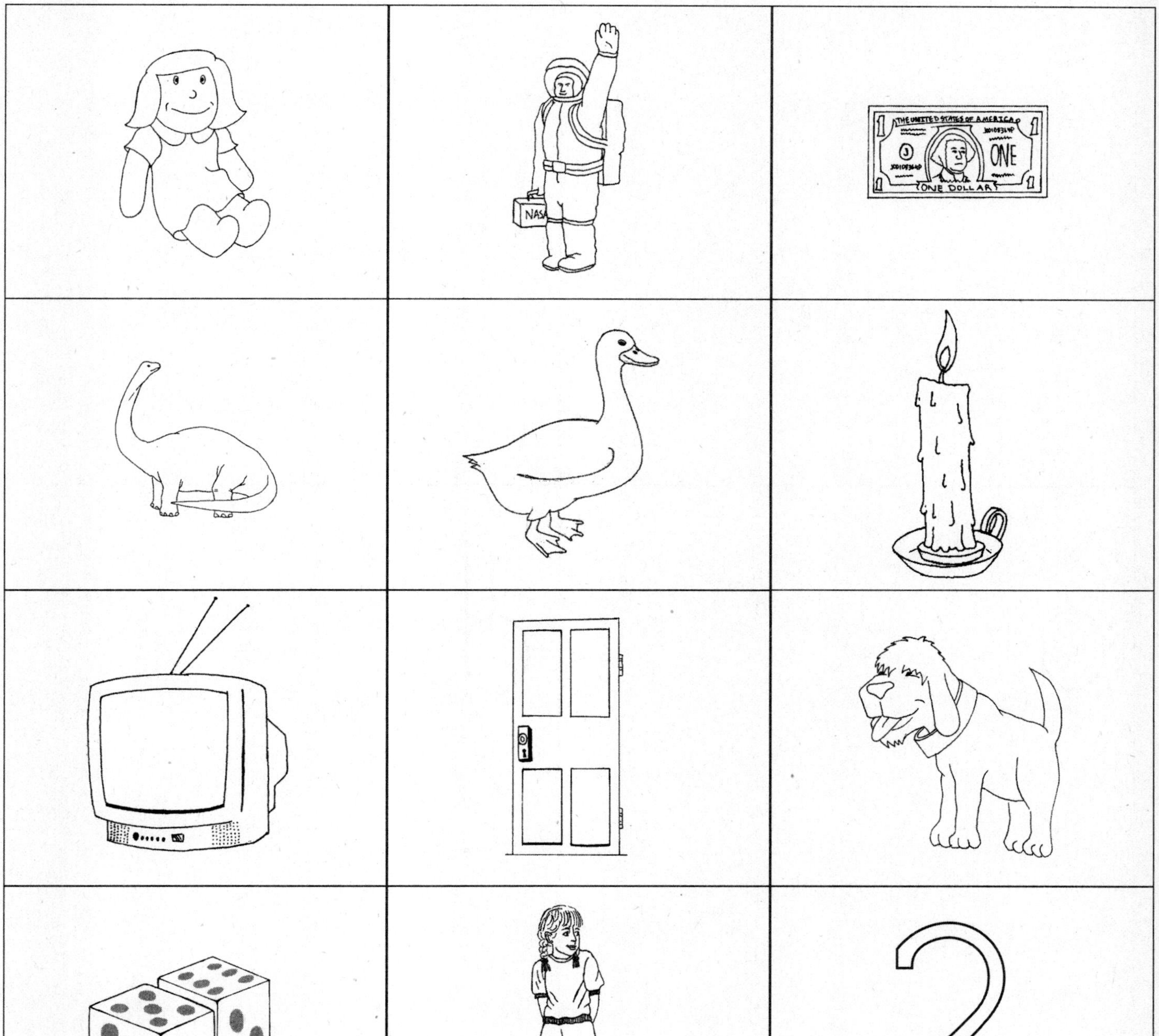

Color the pictures that begin with the sound /*d*/.

c	t
a	d

Draw pictures that begin with each sound.

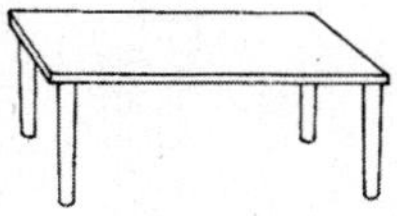

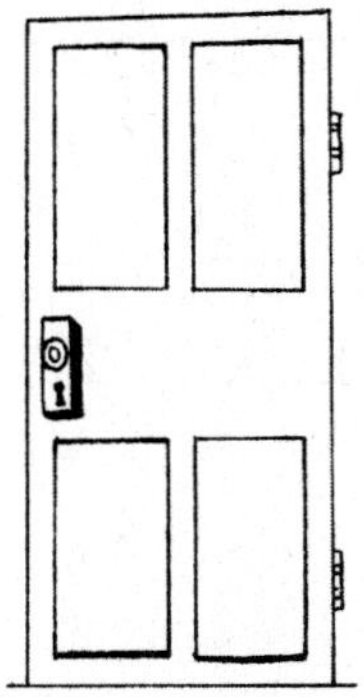

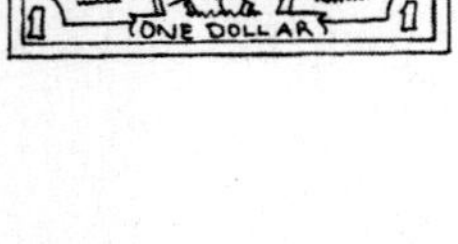

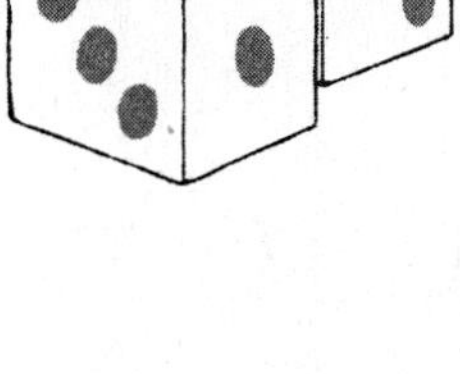

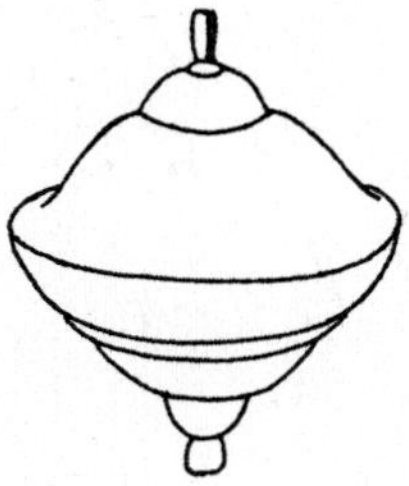

Write the letter *d* next to each picture that begins with the sound /*d*/.

cat

dad

Draw pictures that illustrate each word.

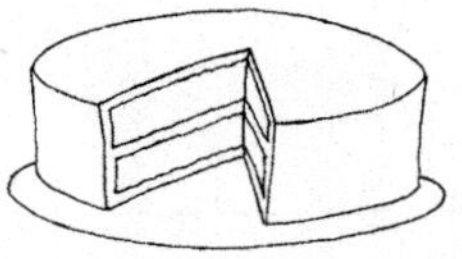

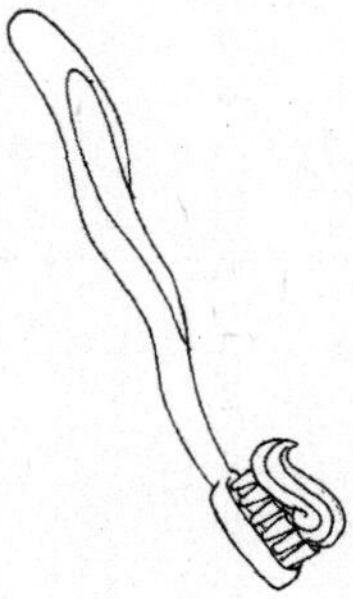

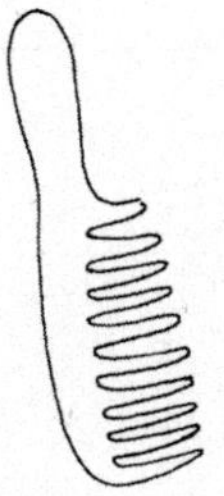

Write the first sound in each word.

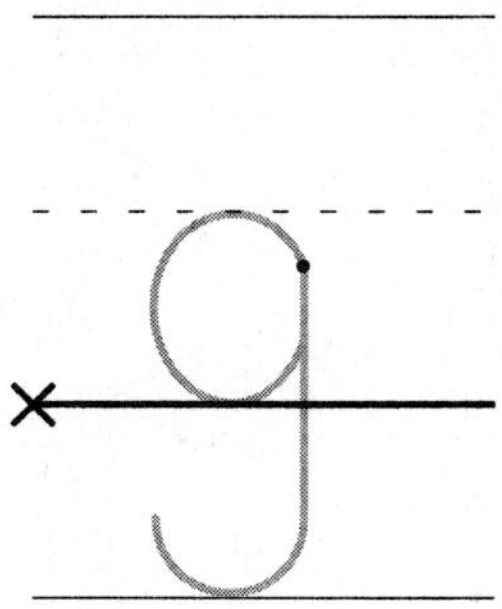

Color the pictures that begin with the sound /*g*/.

a	g
t	d
c	g

Draw pictures that begin with each sound.

g

		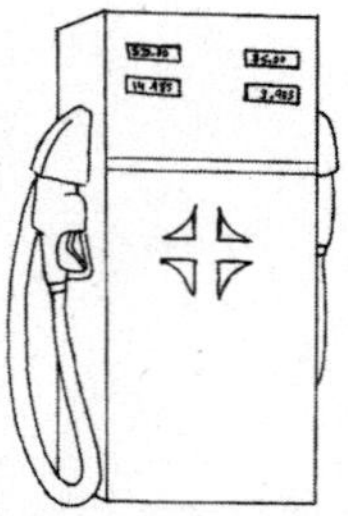

Write the letter *g* next to each picture that begins with the sound /*g*/.

tag

tag

Draw pictures of two of the meanings of *tag* such as a label or the game.

s

Color the pictures that begin with the sound /*s*/.

a	s
g	c
d	t

Draw pictures that begin with each sound.

s

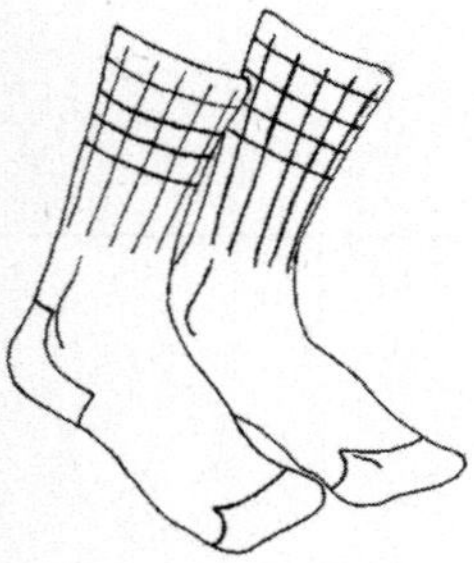

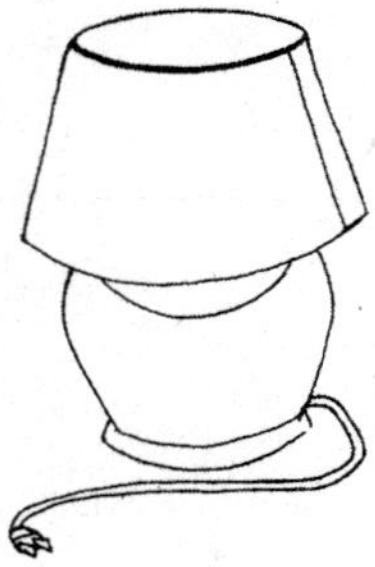

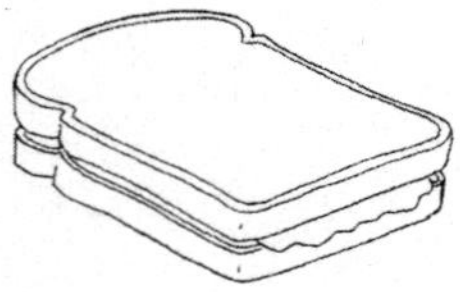

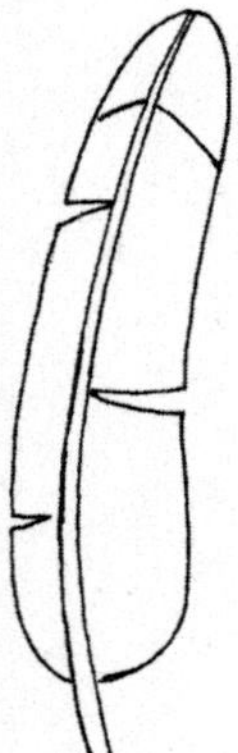

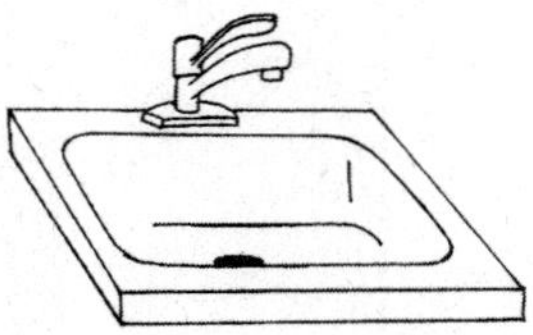

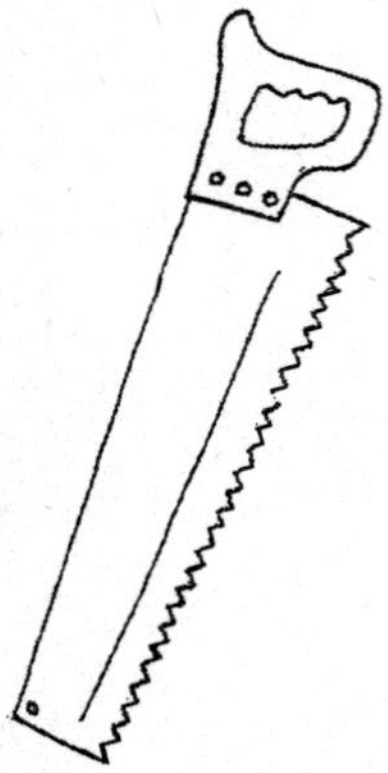

Write the letter *s* next to each picture that begins with the sound /*s*/.

at

cat

dad

sat

sad

tag

gas

Read the list with the children, helping them to blend the sounds into words.

sad cat

sad dad

Draw pictures that illustrate the phrases.

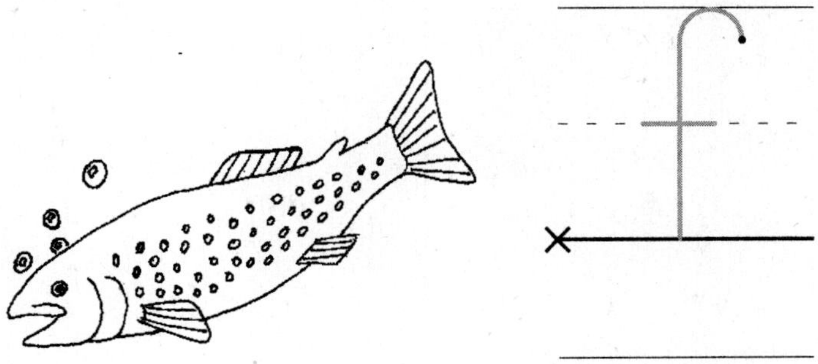

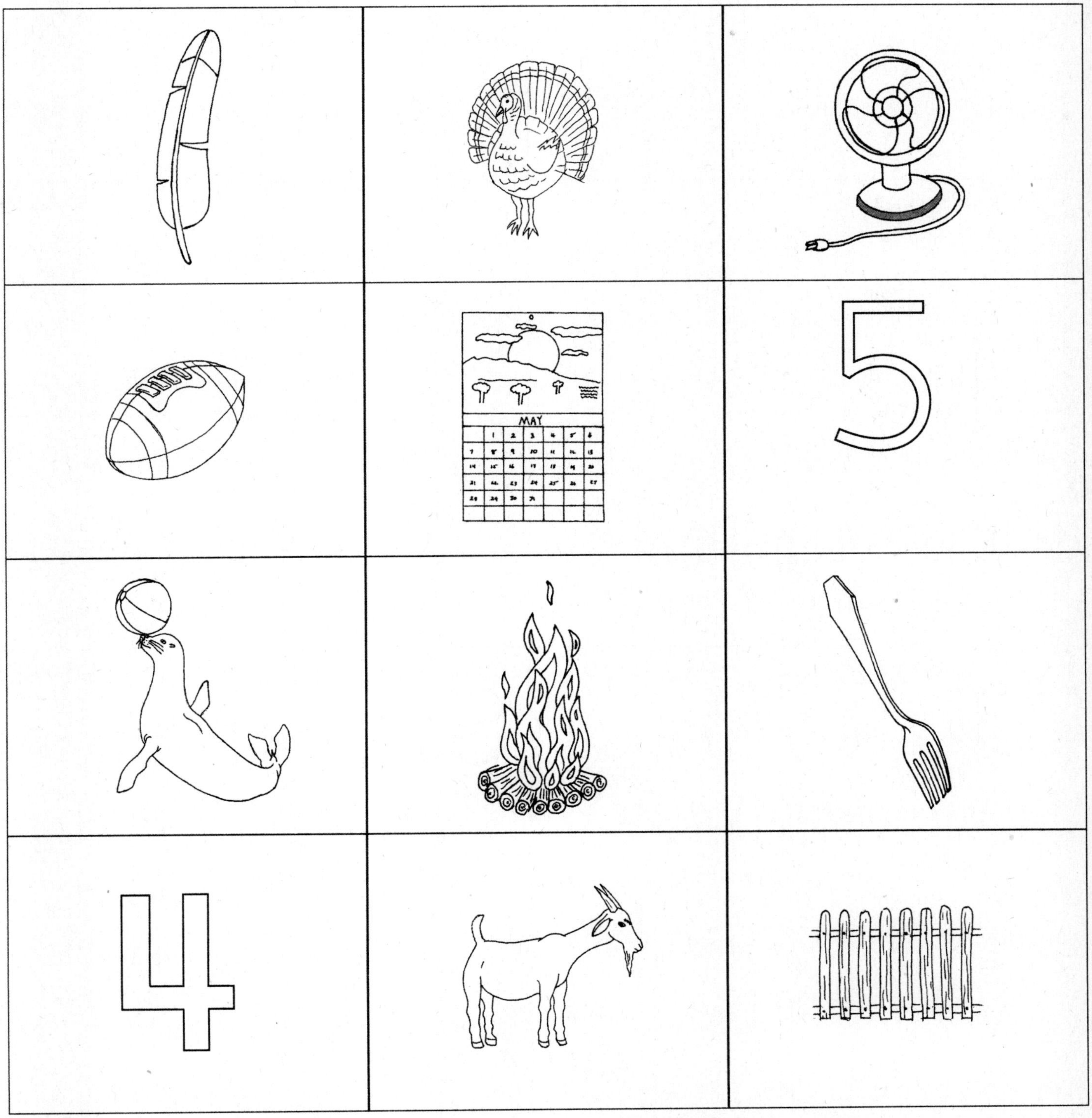

Color the pictures that begin with the sound /*f*/.

f	t
g	d
a	s

Draw pictures that begin with each sound.

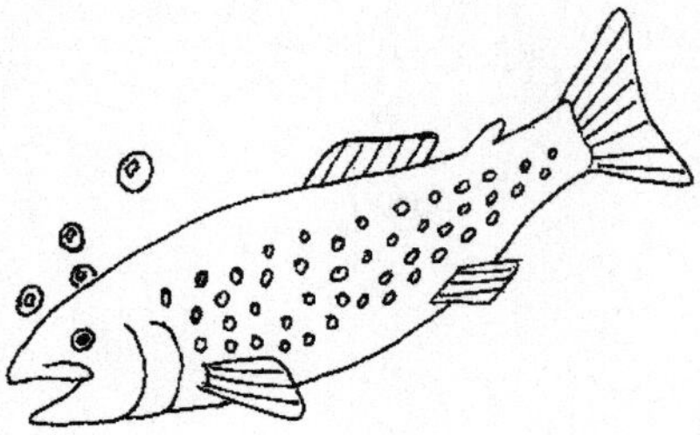

f

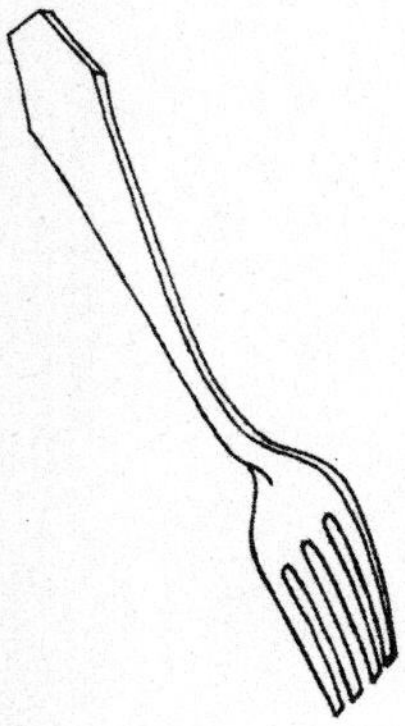

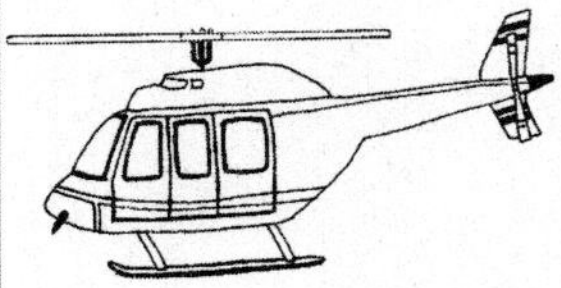

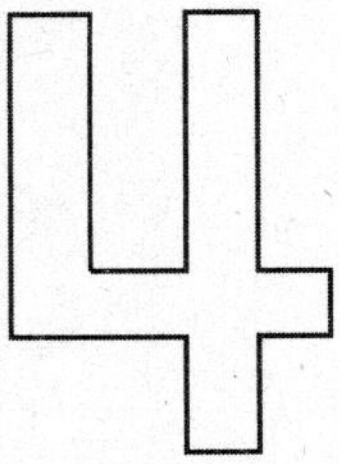

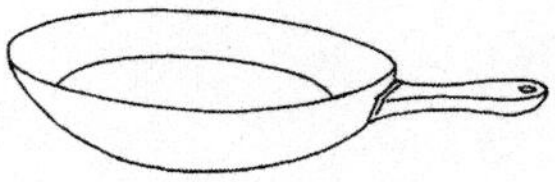

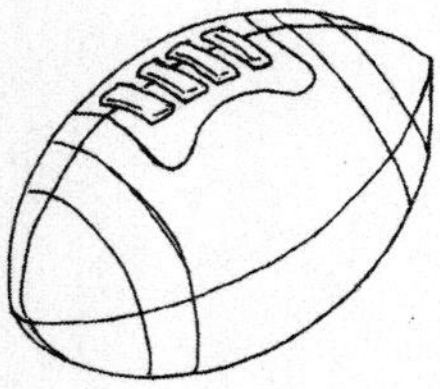

Write the letter *f* next to each picture that begins with the sound /*f*/.

fat	at
cat	gas
dad	tag
sad	sat

sat

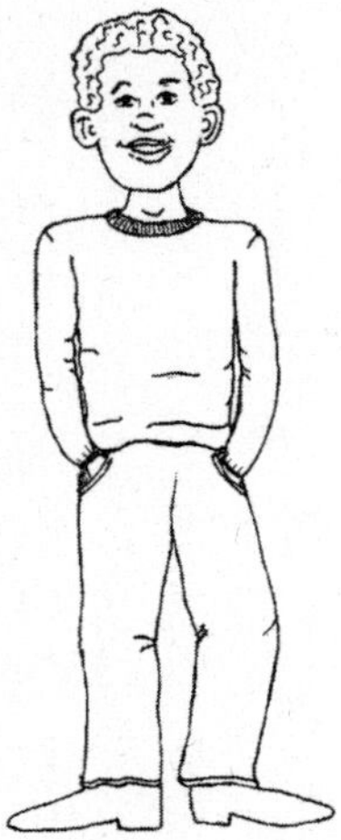

sad

fat

Put an *x* in the box next to the picture that each word describes.

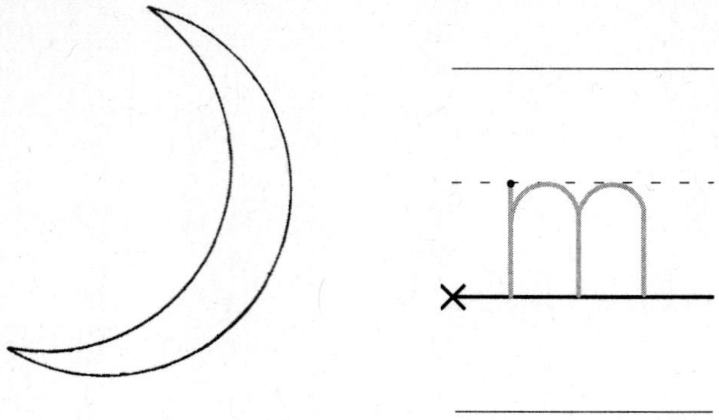

Color the pictures that begin with the sound /*m*/.

m	a
f	d
g	c

Draw pictures that begin with each sound.

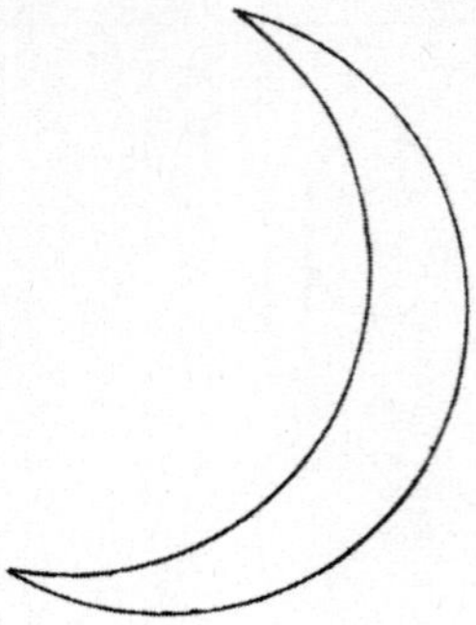

m

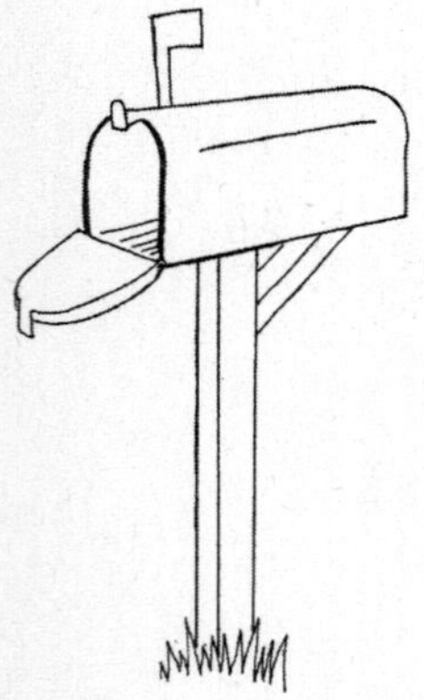

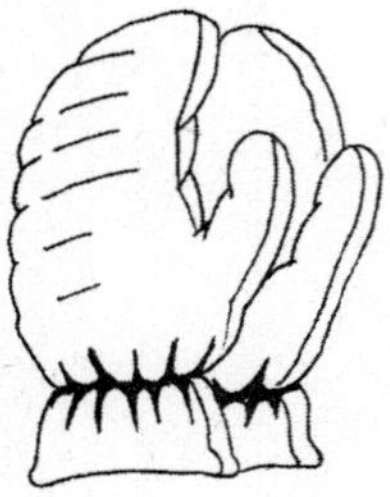

Write the letter *m* next to each picture that begins with the sound /*m*/.

am	sad
mad	at
mat	dad
fat	tag
cat	sat

Read the list with the children.

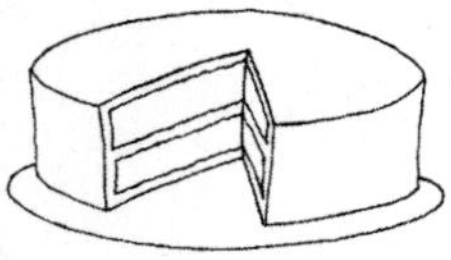

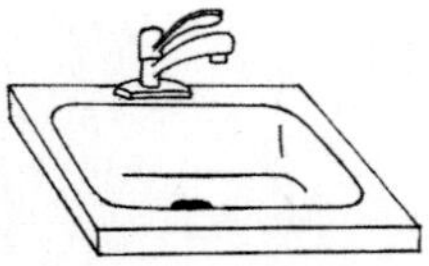

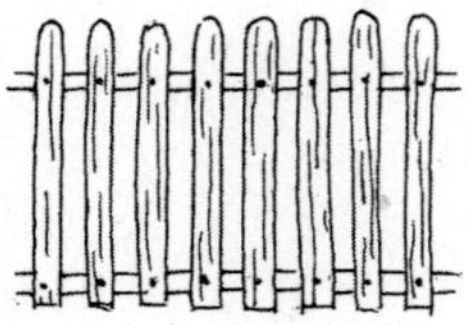

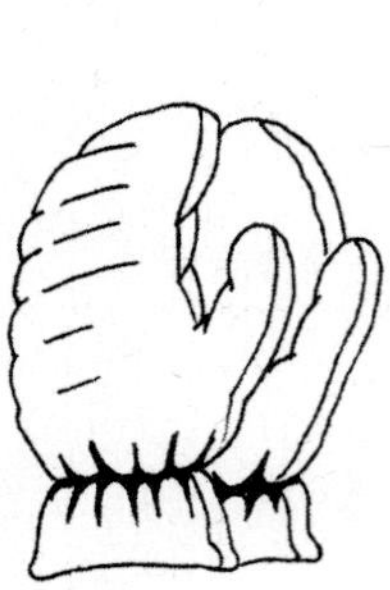

Write the first sound in each word.

sad

mad

Draw pictures that illustrate each word.

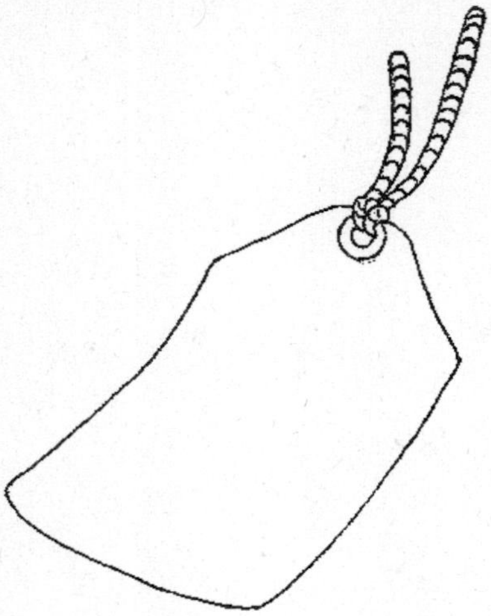

Write the word for each picture.

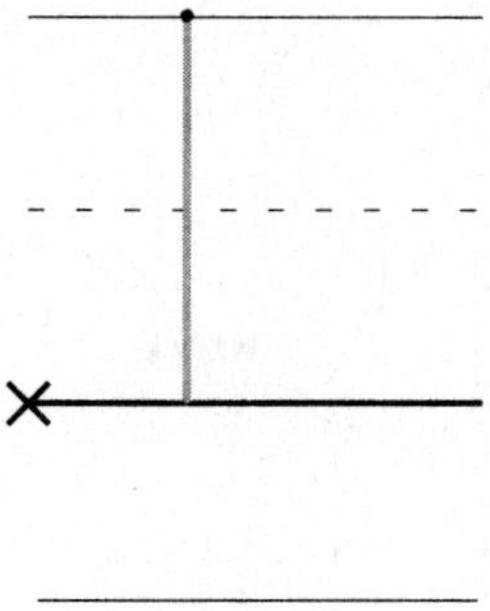

Color the pictures that begin with the sound /*l*/.

l	f
c	s
a	m

Draw pictures that begin with each sound.

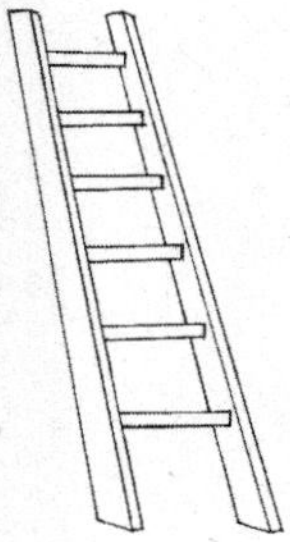

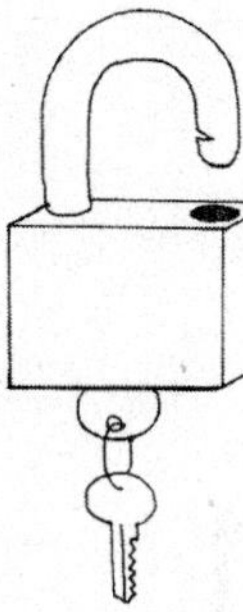

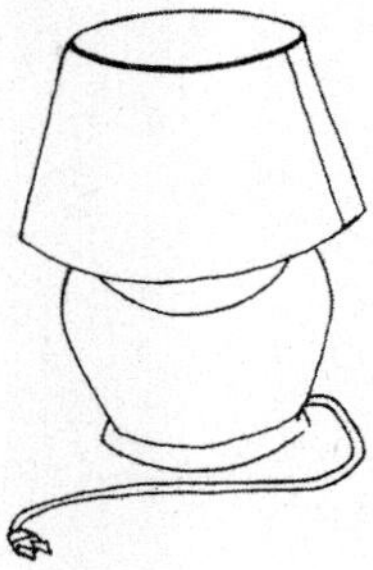

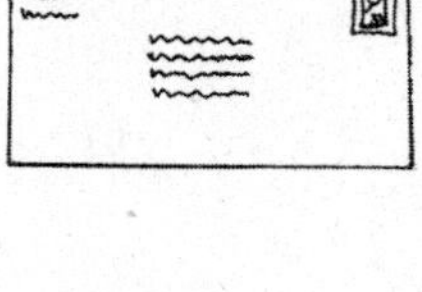

Write the letter *l* next to each picture that begins with the sound /*l*/.

lad	fat	tag
gal	sat	sad
am	dad	at
mad	mat	gas

Read the list with the children.

a mat	a tag
a lad	a fat cat

Draw pictures that illustrate the phrases.

a cat

a tag

a mat

a fat cat

a sad dad

a mad dad

A cat sat.

A dad sat.

Read the text with the children. Underline the two sentences.

A cat sat.

A dad sat.

Draw pictures that illustrate the sentences.

a mad gal

a mat

a fat cat

a tag

a sad gal

Draw a line from the phrase to the appropriate picture.

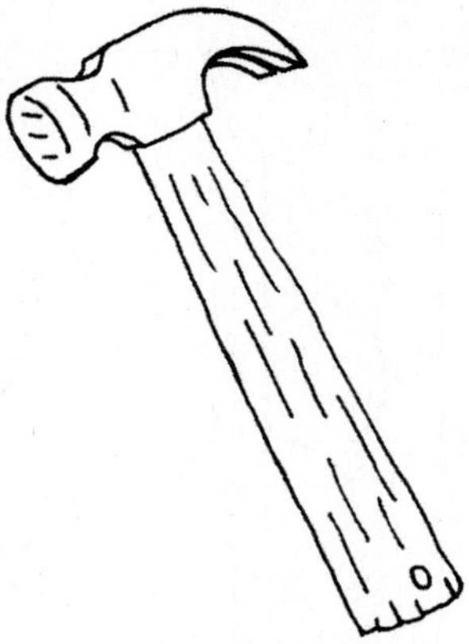

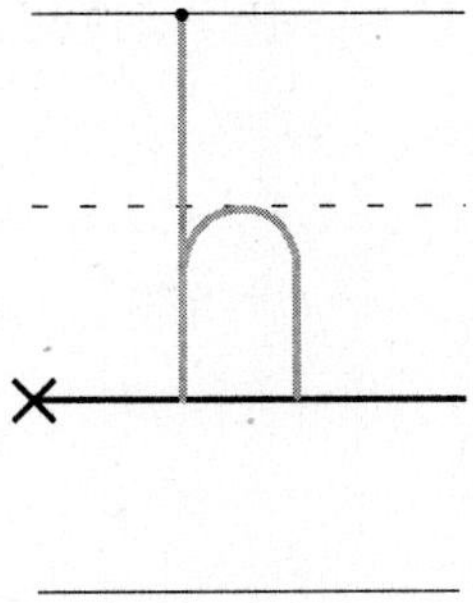

Color the pictures that begin with the sound /*h*/.

h	d
l	f
m	a

Draw pictures that begin with each sound.

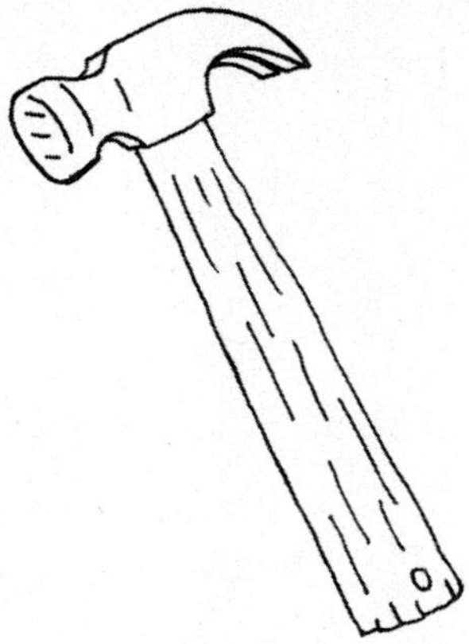

h

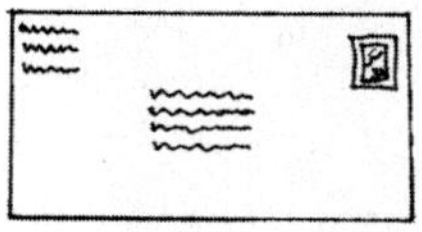

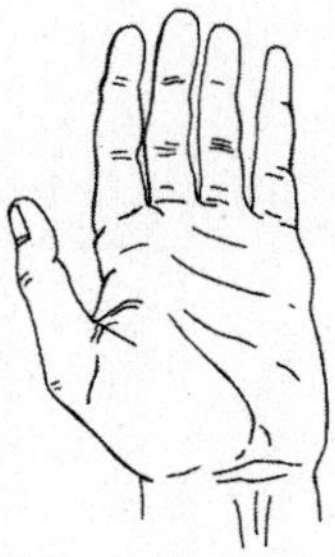

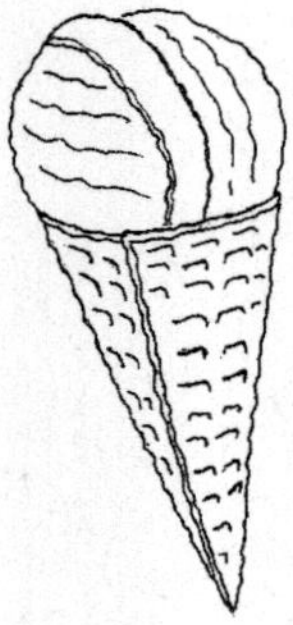

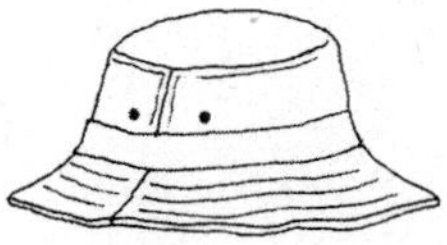

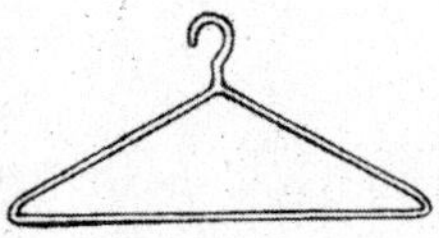

Write the letter *h* next to each picture that begins with the sound /*h*/.

had	mad	lad
hat	am	mat
ham	gal	sad

had a ham

had a hat

a sad gal

Read the list with the children.

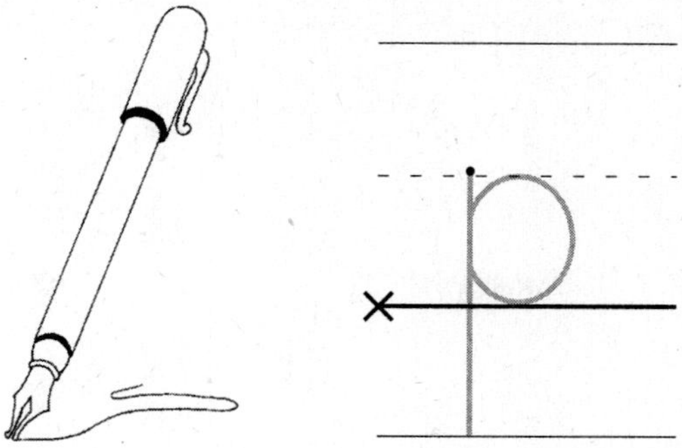

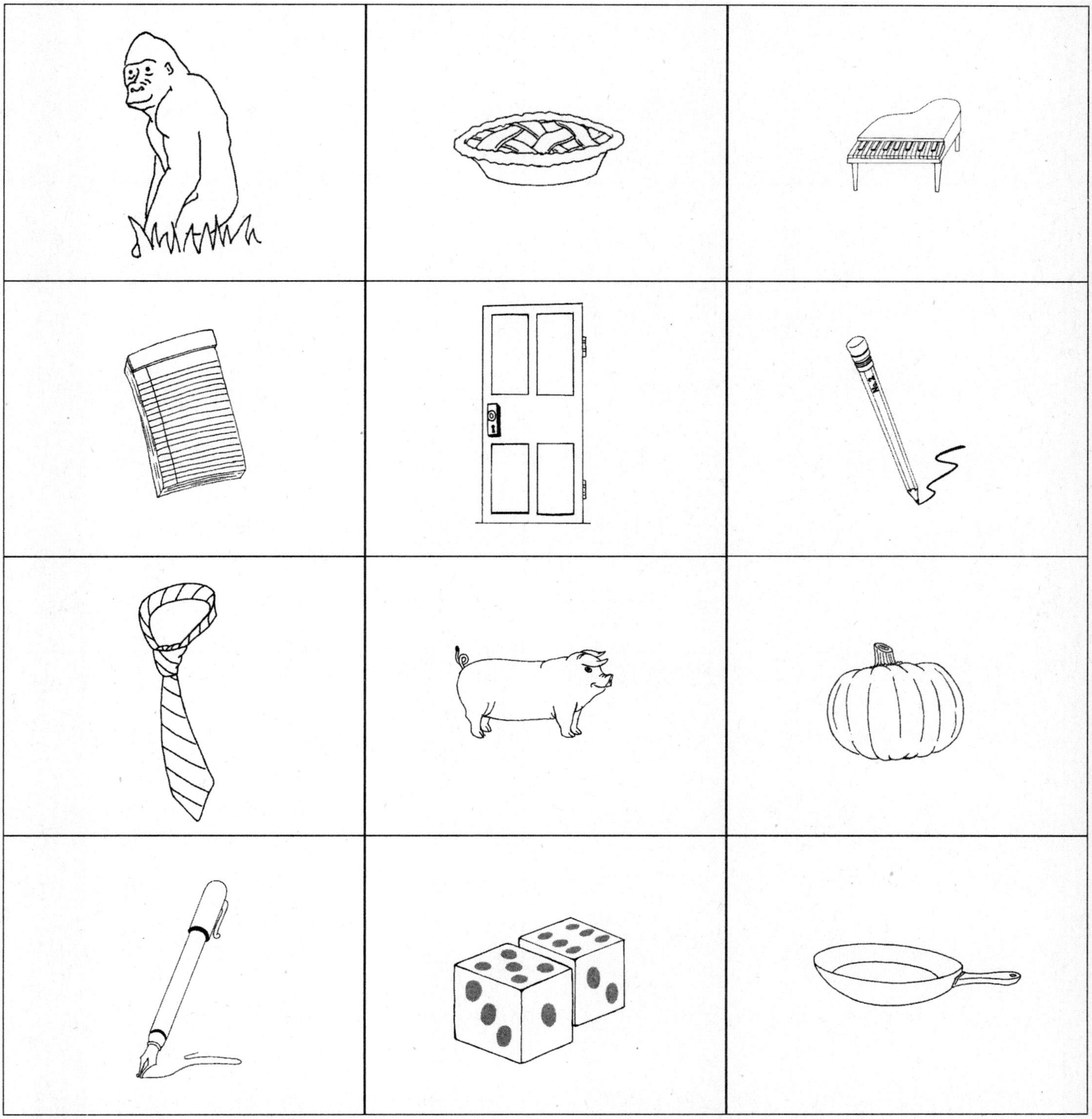

Color the pictures that begin with the sound /*p*/.

p	a
h	l
s	m

Draw pictures that begin with each sound.

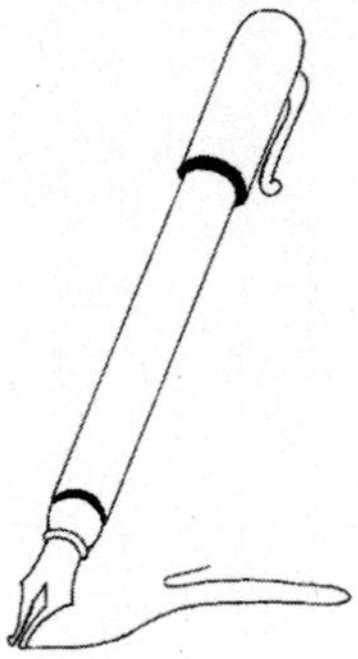

p

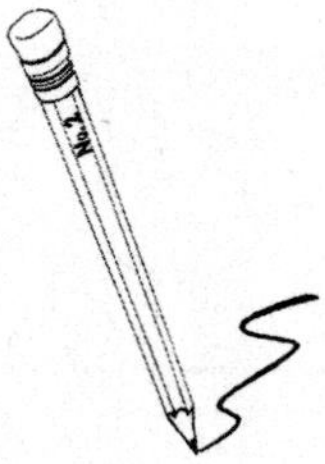

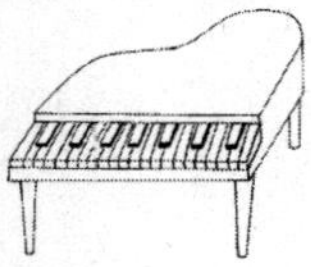

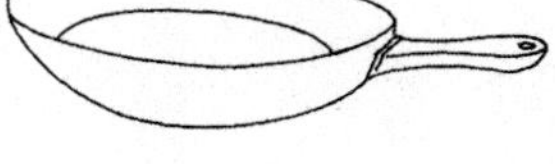

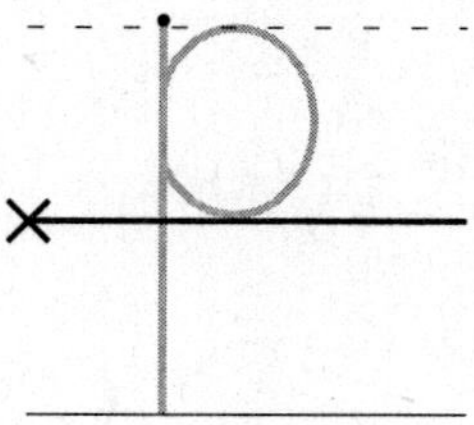

Write the letter *p* next to each picture that begins with the sound /*p*/.

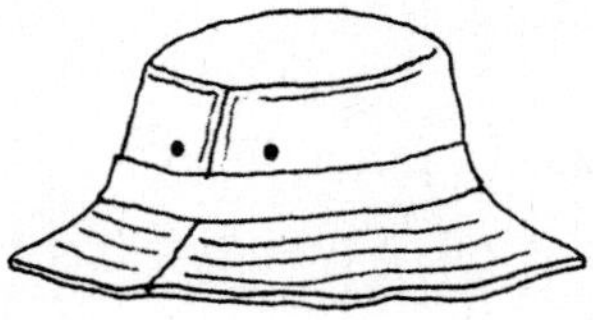

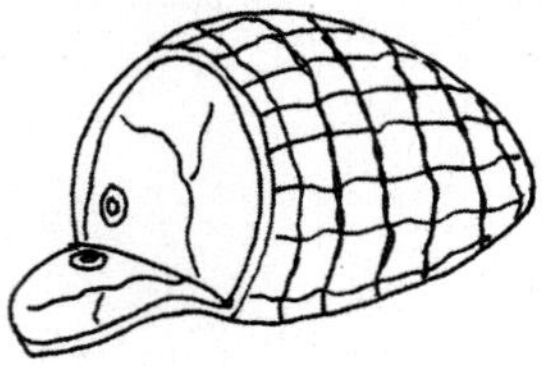

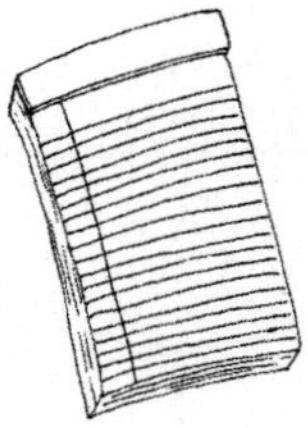

Write the word for each picture.

tap	pat	mat
lap	cap	mad
pal	pad	hat
map	ham	am

a pal

a map

a cap

a pad

Read the list with the children.

a cap

a pad

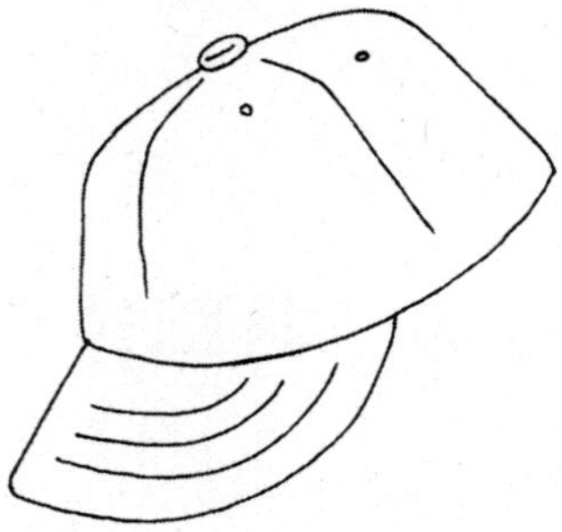

a map

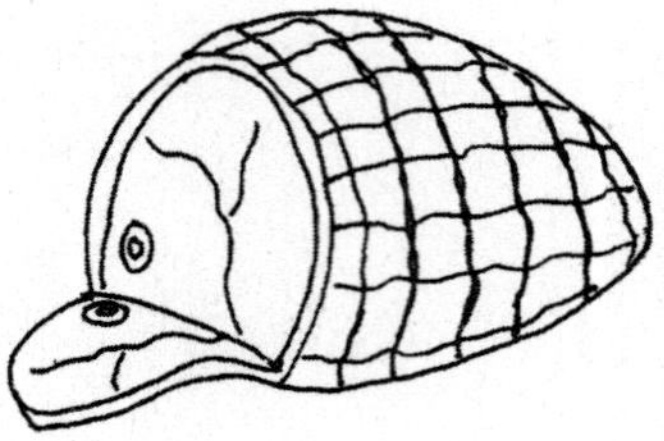

a ham

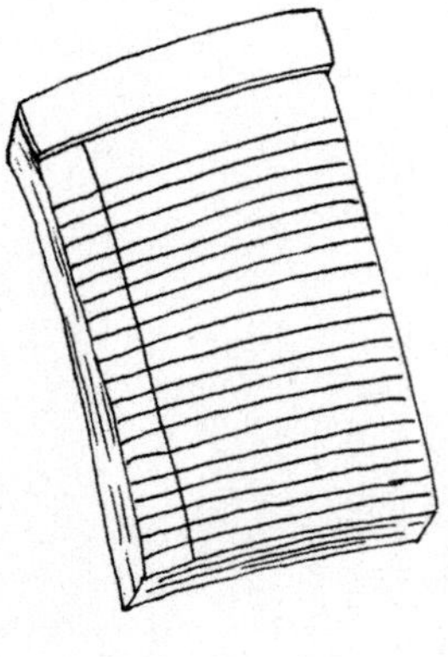

Draw a line from each phrase to the appropriate picture.

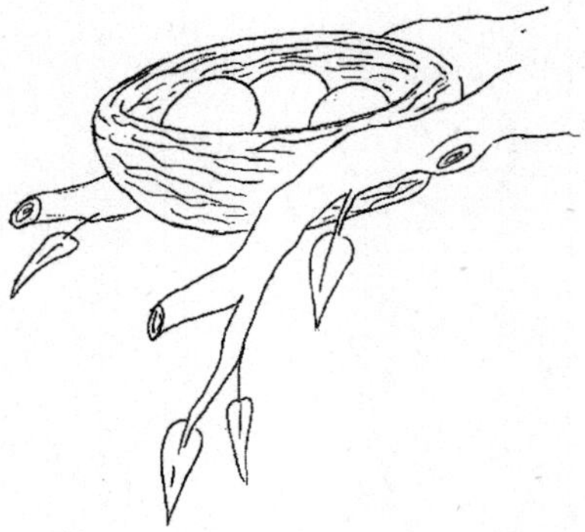

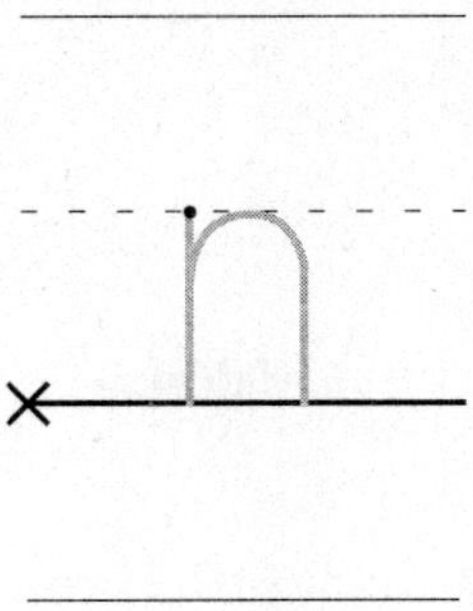

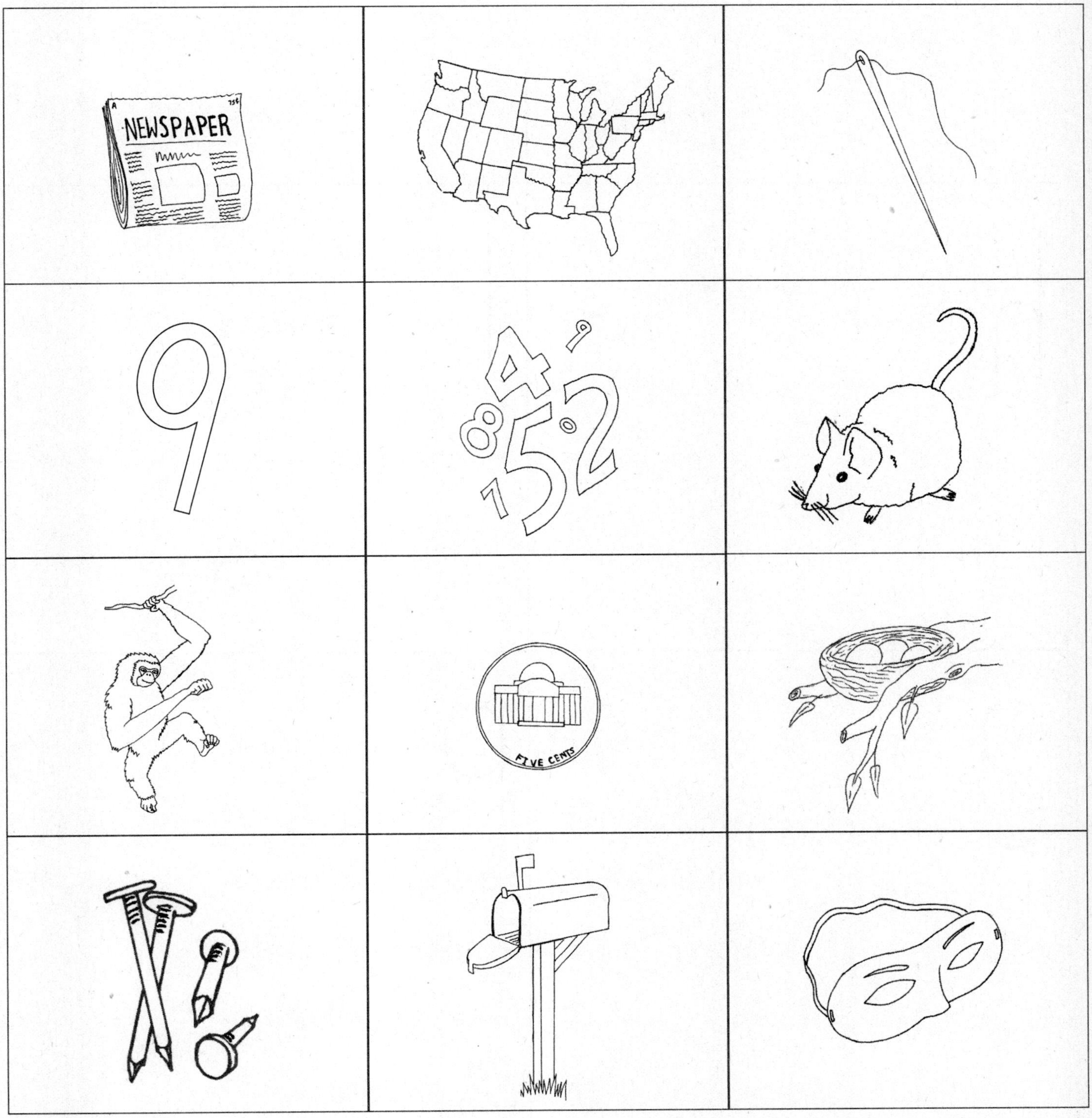

Color the pictures that begin with the sound */n/*.

n	a
h	l
g	p

Draw pictures that begin with each sound.

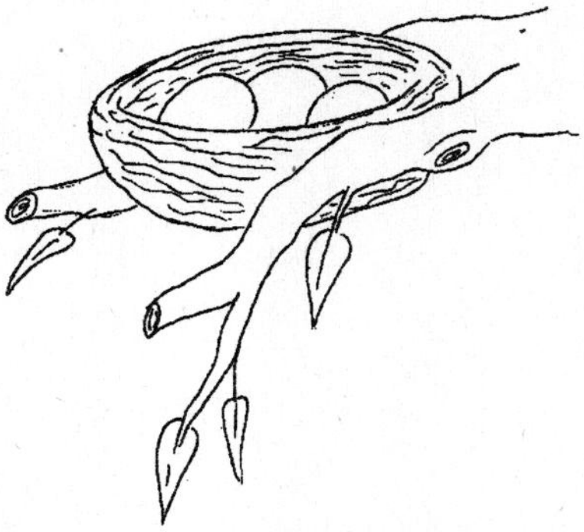

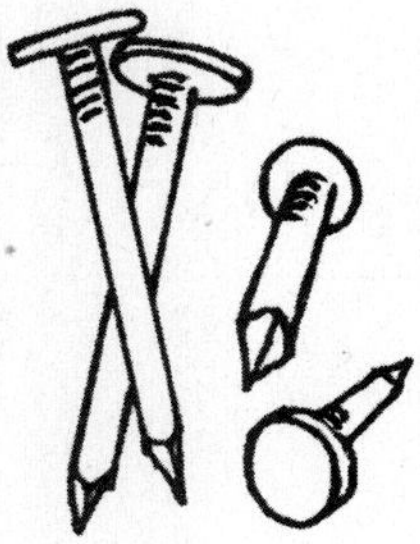

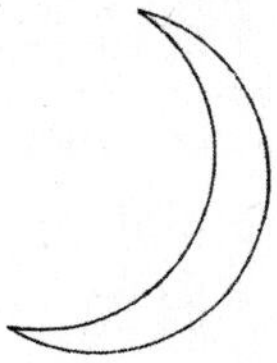

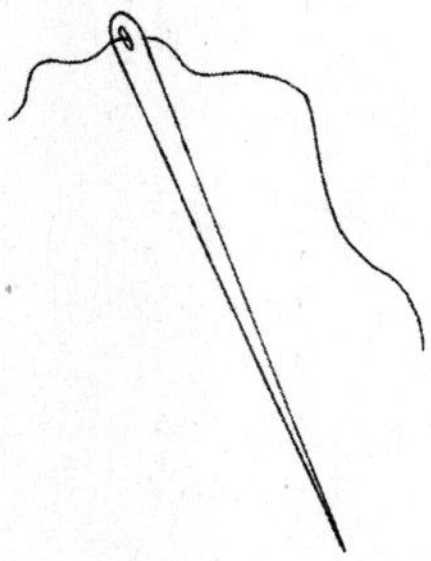

Write the letter *n* next to each picture that begins with the sound /*n*/.

can	tan	tap
man	map	had
nap	fan	and

a tan fan

a gas can

had a nap

a sad man

Read the list with the children.

a cat **and** a hat

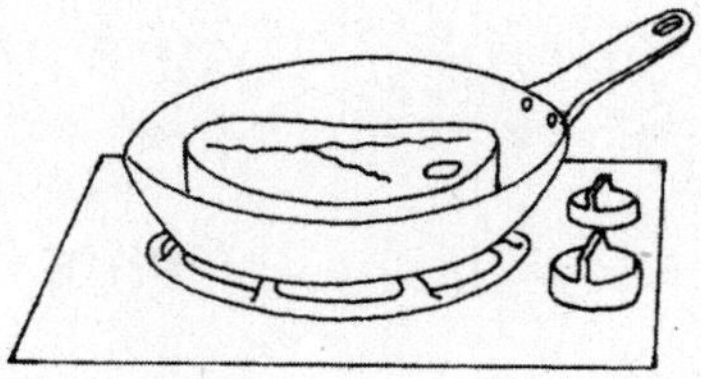

a pan **and** a can

a man **and** a map

a ham **and** a pan

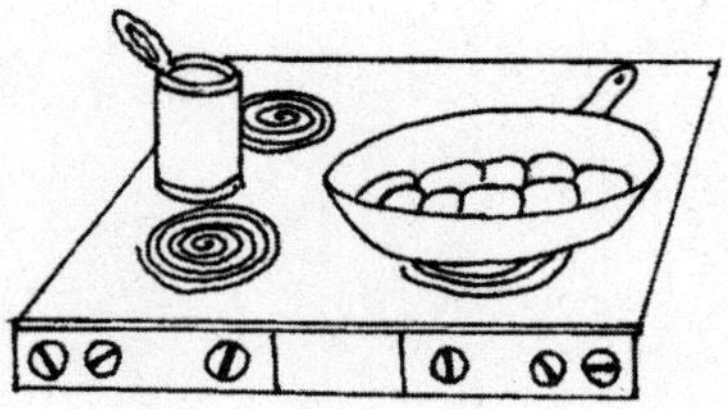

a fat cat **and** a mat

Draw a line from the word to the appropriate picture.

cat	**tap**
can	**tan**
cap	**tag**
mat	**pat**
mad	**pad**
man	**pan**
map	**pal**

Read the list with the children.

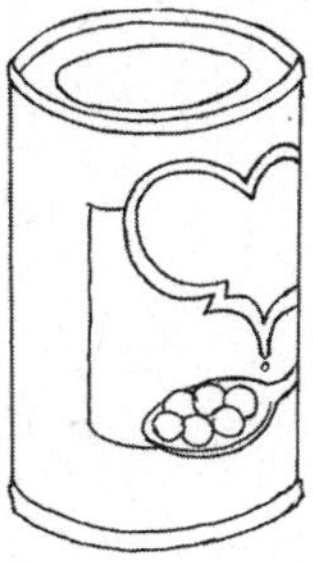

cat

can

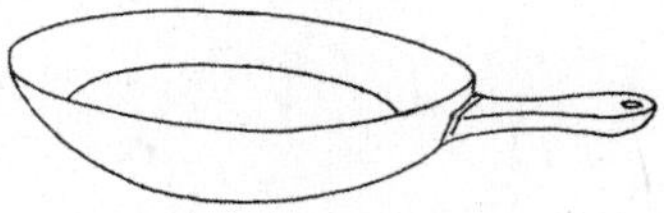

pat

pan

sat

sad

man

map

fan

fat

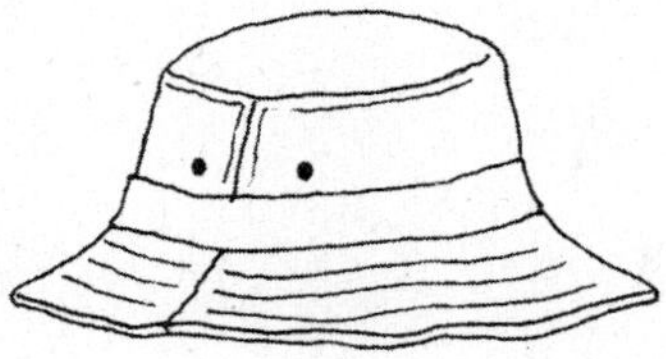

had

hat

mad

mat

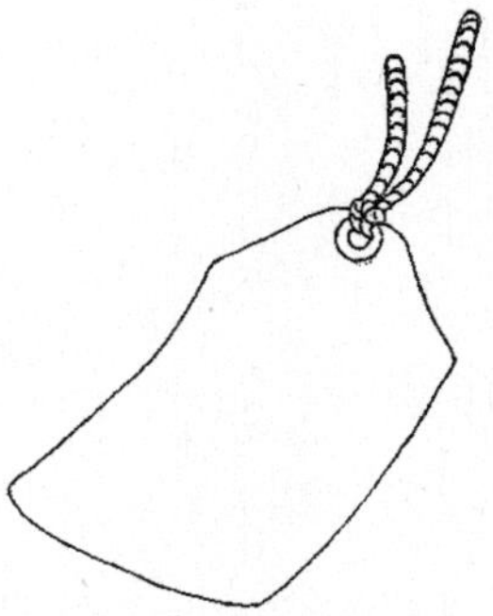

tan

tag

Circle the word for each picture.

I am sad.

I am mad.

Draw pictures that illustrate the sentences.

I am a man.

I am a fat cat.

I had a ham.

I had a nap.

I had a cap.

I had a can.

I can tag dad.

I can pat a cat.

Circle the sentence that matches each picture.

cat	hat	map	cap
cats	hats	maps	caps

1 cat	3 hats
2 cats	1 hat

Draw pictures that illustrate the phrases.

I can nap.
I am Nat.

Nat had a pan.
Nat can nap.

I can pat Nat.
I am mad at Nat.

Circle the sentence that matches each picture.

I am Nat.

I am a cat.

I am a fat cat.

I am Al.

I am a man.

I can nap.

I am Al.

I can pat a cat.

I can pat Nat.